BLACK INVENTORS

BY
KING KI'EL

Copyright 2023 © Urbantoons Black Inventors Illustrated by URBANTOONS Publishing All rights reserved By Sheila Anne Simmons. No part of this publication may be reproduced, distributed, or transmitted in any form or by any means, including photocopying, recording, or other electronic or mechanical methods, without the prior written permission(permission) of the publisher, except in the case of brief quotations bodied in critical reviews and certain other noncommercial uses permitted by copyright law. For permission requests, write to the publisher, addressed "Attention: Permissions Coordinator," Publisher's Note: This is a work of fiction. Names, characters, places, and incidents are a product of the author's imagination. Locales and public names are sometimes used for atmospheric purposes. Any resemblance to actual people, living or dead, businesses, companies, events, institutions, or locales is completely coincidental.

BLACK AUTHORS

Black history is more than just a collection of facts and figures about the accomplishments of black people. It is a vital narrative that encompasses the struggles, triumphs, and contributions of a group of people who have been marginalized and oppressed for centuries. As a black person, I believe that it is essential to hear the stories of our history and culture from the perspective of black authors. When black history is told by black authors, it allows for a deeper

understanding of the complexities and nuances of the black experience. Black authors bring a unique perspective to the telling of our history, one that is rooted in our own lived experiences and understanding of the world. They are able to share the stories of our ancestors in a way that is both honest and authentic, providing a more complete and accurate picture of the past.

Furthermore, black history told by black authors also serves as a powerful tool for self-empowerment and self-affirmation. It allows us to see ourselves reflected in the stories of our ancestors and understand that we are not alone in our struggles. It provides a sense of connection to a rich cultural heritage that can inspire pride and a

sense of belonging.

In addition, black history told by black authors is essential for the education of all people. It helps to promote understanding and empathy for the black experience and to challenge the dominant narratives that have long been used to marginalize and oppress black people. It allows for a more inclusive and accurate portrayal of history, one that acknowledges and celebrates the contributions of

black people.

In conclusion, black history is an essential aspect of understanding the world we live in today, and it is important to hear the stories of our history and culture from the perspective of black authors. Black authors bring a unique and valuable perspective to the telling of our history, serving as powerful tools for self-empowerment and education for all people. Let's acknowledge, appreciate and celebrate the black authors for their contributions in the history of the world.

FORWARD

"This book is the first in a series that will explore the brilliant and innovative contributions of black inventors to science, technology, engineering, and mathematics. From the earliest days of our nation's history, black inventors have been at the forefront of technological advancement, overcoming obstacles and discrimination to make groundbreaking discoveries that have changed the course of human progress.

We will learn about the amazing inventions of Lewis Latimer, who developed an improved method for manufacturing the carbon filaments used in incandescent light bulbs, and Elijah McCoy, who patented more than 50 inventions, including the automatic lubrication system that bears his name. We will also discover the pioneering work of George Washington Carver, who developed hundreds of uses for peanuts and sweet potatoes, and Dr. Percy Julian, who synthesized hormones and steroids, including the first synthetic cortisone.

These inventors, and many others featured in this series, have not only left a lasting legacy in their respective fields but also have helped pave the way for future generations of black inventors. They have inspired us to reach for the stars and to never give up on our dreams. This series is a must-read for anyone interested in the history of innovation and the powerful impact of black inventors on our world."

George Washington Carver

Converting Peanuts Into a Variety of Useful Products

1864 - 1943

George Washington Carver was an African American inventor and scientist from Missouri. He was born in 1864 and is known for his contributions to agriculture and plant science.

One of George's most significant inventions was a process for converting peanuts into a variety of useful products, including peanut oil, peanut butter, and a range of other food and non-food products. This was an important innovation, as it helped to increase the economic value of peanuts and made them an important crop in the United States. George's process was widely used by farmers and manufacturers around the world.

In addition to his work on peanuts, George also made significant contributions to the field of agriculture through his research and development of other innovative techniques and technologies. He was a pioneer in the use of crop rotation and soil conservation methods, which helped to improve the productivity and sustainability of farms.

George Washington Carver was an important figure in the history of agriculture and his inventions had a lasting impact on the world. They helped to increase the economic value of peanuts and improve the productivity and sustainability of farms, and they continue to be used today in a variety of applications.

"Education is the key to unlock the golden door of freedom." - George Washington Carver

Benjamin Banneker

Wood Clock and and Banneker's Almanac

1731 - 1806

Benjamin Banneker was an African American inventor and scientist from Maryland. He was born in 1731 and is known for his contributions to astronomy, mathematics, and engineering.

One of Benjamin's most notable inventions was the wooden clock. This clock was the first of its kind to be made entirely out of wood, and it was so accurate that it kept time for over 50 years! Benjamin's clock was a marvel of engineering and helped to pave the way for the development of more advanced clocks and timekeeping devices.

In addition to his work on the wooden clock, Benjamin was also a skilled astronomer and mathematician. He used his knowledge of these subjects to create a series of almanacs, which were popular guides to the movements of the stars and planets.

Today, Benjamin Banneker is remembered as a pioneer in the fields of astronomy and engineering. His inventions and discoveries have had a lasting impact on the world and continue to inspire people of all ages to pursue their passions and achieve great things.

"Presumption should never make us neglect that which appears easy to us, nor despair make us lose courage at the sight of difficulties." –Benjamin Banneker.

Alice H. Parker
Heating and Ventilation
1895 - 1920

Alice H. Parker was an African American inventor from New Jersey. She was born in 1895 and became known for her work in heating and ventilation.

One of Alice's most notable inventions was the gas heating furnace. This furnace allowed people to heat their homes using natural gas, which was more efficient and cost-effective than other heating methods at the time. Alice's furnace was patented in 1919 and quickly became popular in homes across the country.

Alice's invention had a huge impact on the world, as it made it much easier and more affordable for people to heat their homes. This was especially important during the cold winter months when people needed to keep warm.

Today, Alice's gas heating furnace is still used in many homes and buildings around the world. It is a testament to her ingenuity and determination to make the world a better place.

Alice H. Parker was an amazing inventor who used her talents to improve the lives of people everywhere. She is a role model for children everywhere, showing them that they too can make a difference in the world through hard work and determination.

Garret Augustus Morgan

Transportation and War

1877 - 1963

Garret Augustus Morgan was an African American inventor and entrepreneur from Ohio. He was born in 1877 and is known for his contributions to the field of safety and transportation.

One of Garret's most significant achievements was the development of a traffic signal, which he invented in 1923. The traffic signal was an important innovation, as it helped to improve the safety and efficiency of traffic flow in urban areas. Garret's traffic signal was the first to use a "stop and go" system, which is still in use today.

In addition to his work on the traffic signal, Garret also made significant contributions to the field of safety through the development of a gas mask, which he invented in 1914. The gas mask was an important innovation, as it helped to protect people from inhaling harmful gases during World War I.

Garret Augustus Morgan was an important figure in the history of engineering and his achievements had a lasting impact on the world. His work helped to improve the safety and efficiency of transportation and continues to be recognized as an important milestone in the history of science and technology.

"If you can be the best, then why not try to be the best?" -- Garrett Morgan

Granville T. Woods

Fields of Electricity, Transportation and Telecommunications

1856 – 1910

Granville T. Woods was a brilliant inventor and engineer from Columbus, Ohio. Even though he was orphaned at a young age, he worked hard and became an expert in the field of electrical engineering. One day, Granville had a brilliant idea - he wanted to find a way for people to send messages over long distances using electricity. So, he worked hard and developed the "induction telegraph," which allowed people to communicate faster by using their voices over telegraph wires. It was such an important invention that even Thomas Edison challenged Granville's patent with a lawsuit, but Granville won and proved his genius to the world.

But that wasn't the only amazing invention Granville created. He also developed a device called the "troller," which was a special wheel that allowed street cars, also known as "trolleys," to collect electricity from overhead wires. This made it easier for people to travel and made transportation more efficient. Granville's inventions had a huge impact on the world and helped to improve the efficiency and reliability of electrical systems. He inspired many other inventors and engineers to pursue careers in the field of electrical engineering, and his legacy as an inventor and engineer is a great source of inspiration for all children. Granville's dedication to his work and his commitment to making the world a better place through innovation are valuable lessons for us all.

"SOME men are born great; some have greatness thrust upon them; and some achieve greatness." - Granville T. Woods

Lewis Latimer
Field of Electricity
1848 - 1928

Lewis Latimer was an African American inventor and engineer from Massachusetts. He was born in 1848 and is known for his contributions to the field of electricity.

One of Lewis' most significant inventions was a process for manufacturing high-quality carbon filaments for use in light bulbs. This process, which involved creating a uniform, high-grade carbon thread that could be used as a filament in light bulbs, was an important innovation, as it helped to improve the efficiency and reliability of light bulbs. Lewis' process was widely used by manufacturers around the world and helped to make electric lighting more affordable and widely available.

In addition to his work on carbon filaments, Lewis also made significant contributions to the field of electricity through his research and development of other innovative technologies. He was a pioneer in the use of electric lighting and helped to lay the foundation for many of the electrical technologies that we use today.

Lewis Latimer was an important figure in the history of electricity and his inventions had a lasting impact on the world. They helped to improve the efficiency and reliability of light bulbs and continue to be used in a variety of applications.

"We create our future, by well improving present opportunities: however few and small they are." - Lewis Latimer

Percy Julian

Field of Medicine

1899 - 1975

Percy Julian was an African American inventor and chemist from Alabama. He was born in 1899 and is known for his contributions to the field of medicine.

One of Percy's most significant inventions was a drug that could be used to treat glaucoma, a condition that causes damage to the optic nerve and can lead to blindness. This drug, known as "Pilocarpine," was an important innovation, as it was the first effective treatment for glaucoma and helped to save the sight of many people around the world. Percy's Pilocarpine was patented in 1950.

In addition to his work on Pilocarpine, Percy also made significant contributions to the field of medicine through his research and development of other innovative drugs and treatments.

Percy Julian was an important figure in the history of medicine and his inventions had a lasting impact on the world. They helped to improve the treatment of glaucoma and continue to be used in a variety of medical applications.

Jan Matzeliger

Field of Shoe Manufacturing

1852 - 1889

Jan Matzeliger was an African American inventor and engineer from Suriname. He was born in 1852 and is known for his contributions to the field of shoe manufacturing.

One of Jan's most significant inventions was a machine that could attach the upper part of a shoe to its sole more quickly and efficiently than any other method at the time. This machine, known as the "lasting machine," was an important innovation, as it helped to revolutionize the shoe industry and made it possible to produce shoes much more cheaply and quickly than before. Jan's lasting machine was patented in 1883 and was widely used by shoe manufacturers around the world.

In addition to his work on the lasting machine, Jan also made significant contributions to the field of shoe manufacturing through his research and development of other innovative technologies.

Jan Matzeliger was an important figure in the history of shoe manufacturing and his inventions had a lasting impact on the world. They helped to revolutionize the shoe industry and continue to be used in the production of shoes today.

James West

Field of Acoustics

1931 -

James West was an African American inventor and engineer from Delaware. He was born in 1931 and is known for his contributions to the field of acoustics.

One of James' most significant inventions was the electret microphone, which is a type of microphone that uses a thin, flexible diaphragm to convert sound waves into an electrical signal. This was an important innovation, as it made microphones more sensitive and accurate, and it is still used today in a wide range of applications, including telephones, microphones, and speakers. James' electret microphone was patented in 1962 and was widely used in the electronics industry.

In addition to his work on the electret microphone, James also made significant contributions to the field of acoustics through his research and development of other innovative technologies.

James West was an important figure in the history of acoustics and his inventions had a lasting impact on the world. They helped to improve the sensitivity and accuracy of microphones and continue to be used in a variety of applications today.

Andrew Beard

Field of Railroads

1849 - 1921

Andrew Beard was an African American inventor from Alabama. He was born in 1849 and is best known for his work in the field of railroads.

One of Andrew's most significant inventions was the rotary snowplow, which was used to clear snow from railroad tracks. This was a crucial invention, as it allowed trains to continue operating during the winter months when snow and ice could cause delays or accidents. Andrew's rotary snowplow was patented in 1887 and was widely used by railroads across the United States.

In addition to his work on the rotary snowplow, Andrew also invented a number of other devices that were used in the railroad industry, including a locomotive truck and a fish-plate for rail joints.

Andrew Beard was an important figure in the history of transportation and his inventions had a significant impact on the world. They helped to make rail travel safer and more efficient, and they are still used in the railroad industry today.

Benjamin Bradley

Field of Steam Engineering

1842 - 1904

Benjamin Bradley was an African American inventor from Maryland. He was born in 1842 and is known for his work in the field of steam engineering.

One of Benjamin's most significant inventions was the steam boiler furnace, which was used to generate steam for powering machinery and other applications. This was an important innovation, as it made it much easier and more efficient to generate steam, which was a key source of energy in the industrial era. Benjamin's steam boiler furnace was patented in 1871 and was widely used in factories and other industrial settings around the world.

In addition to his work on the steam boiler furnace, Benjamin also made significant contributions to the field of steam engineering through his research and development of other innovative devices and technologies.

Benjamin Bradley was an important figure in the history of steam engineering and his inventions had a lasting impact on the world. They helped to make steam power more reliable and efficient, and they are still used today in a variety of applications.

Ernest Everett

Field of Biology and Physiology

1883 - 1941

Ernest Everett Just was an African American inventor and scientist from South Carolina. He was born in 1883 and is known for his contributions to the fields of biology and physiology.

One of Ernest's most significant contributions was his work on the fertilization of eggs in sea urchins. His research helped to shed light on the fundamental process of fertilization and led to a greater understanding of how organisms reproduce. Ernest's work was groundbreaking and had a lasting impact on the field of biology.

In addition to his research on fertilization, Ernest also made significant contributions to the field of physiology through his work on the effects of salt on cells. He developed a method for studying cells in a laboratory setting, which helped to advance our understanding of how cells function and how they are affected by different factors.

Ernest Everett Just was a pioneer in the fields of biology and physiology and his contributions have had a lasting impact on the world. His work helped to advance our understanding of the fundamental processes of life and continues to inspire scientists today.

Frederick D. Jones
Field of Refrigeration
1893 - 1961

Frederick D. Jones was an African American inventor from Ohio. He was born in 1893 and is known for his work in the field of refrigeration.

One of Frederick's most significant inventions was a refrigeration system for long-haul trucks, which allowed perishable goods to be transported over long distances without spoiling. This was a crucial innovation, as it made it possible for trucking companies to transport fresh produce and other perishable items across the country. Frederick's refrigeration system was patented in 1949 and was widely used by trucking companies.

In addition to his work on the refrigeration system for long-haul trucks, Frederick also made significant contributions to the field of refrigeration through his research and development of other innovative devices and technologies.

Bessie Blount Griffin
Field of Medicine
1914 - 2009

Bessie Blount Griffin was an African American inventor and nurse from Virginia. She was born in 1914 and is known for her contributions to the field of medicine.

One of Bessie's most significant inventions was a device that allowed people with disabilities to feed themselves. This device, known as the "Portable Attachable Grasping and Manipulating Apparatus," was designed to help people who had lost the use of their hands or arms due to injury or illness. Bessie's device was patented in 1951 and was widely used in hospitals and rehabilitation centers around the world.

In addition to her work on the "Portable Attachable Grasping and Manipulating Apparatus," Bessie also made significant contributions to the field of medicine through her research and development of other innovative devices and technologies.

Bessie Blount Griffin was an important figure in the history of medicine and her inventions had a lasting impact on the world. They helped to improve the lives of people with disabilities and continue to be used in hospitals and rehabilitation centers today.

Henry Sampson
Field of Nuclear Engineering
1934 - 2015

Henry Sampson was an African American inventor and engineer from Mississippi. He was born in 1934 and is known for his contributions to the field of nuclear engineering.

One of Henry's most significant inventions was a device that could generate electricity from the radioactive decay of certain elements. This device, known as a "gamma-electric cell," was an important innovation, as it allowed for the generation of electricity in places where there was no access to traditional sources of power. Henry's gamma-electric cell was patented in 1971 and was widely used in remote locations around the world.

In addition to his work on the gamma-electric cell, Henry also made significant contributions to the field of nuclear engineering through his research and development of other innovative technologies.

Henry Sampson was an important figure in the history of nuclear engineering and his inventions had a lasting impact on the world. They helped to provide access to electricity in remote locations and continue to be used in a variety of applications.

George Crum
Field of Food Science and the Development of New Culinary Techniques
1824 - 1914

George Crum was an African American chef and inventor from New York. He was born in 1824 and is known for his contributions to the field of food science and the development of new culinary techniques.

One of George's most significant achievements was the creation of a new type of snack food called "potato chips," which he invented in the 1850s. Potato chips were a revolutionary snack, as they were thin, crispy, and easy to eat. Before the invention of potato chips, people had to rely on other, less convenient snacks, such as apples or bread.

In addition to his work on potato chips, George was also known for his culinary skills and his ability to create delicious dishes using innovative techniques. He was a pioneer in the field of food science and his contributions have had a lasting impact on the world of cooking.

George Crum was an important figure in the history of food science and his achievements had a lasting impact on the world. His work helped to revolutionize the way we think about snacks and continues to be recognized as an important milestone in the history of culinary arts.

David N. Crosthwait Jr.

Field of Heating, Ventilation, and Air Conditioning (HVAC) System.

1898 - 1976

David N. Crosthwait Jr. was an African American inventor and engineer from Tennessee. He was born in 1898 and is known for his contributions to the field of heating, ventilation, and air conditioning (HVAC) systems.

One of David's most significant achievements was the development of a new type of thermostat, which he invented in the 1930s. The thermostat was an important innovation, as it helped to control the temperature in buildings more accurately and efficiently.

In addition to his work on thermostats, David also made significant contributions to the field of HVAC systems through his research and development of other innovative technologies and techniques. He has been awarded numerous patents for his inventions and has been recognized for his contributions to the field of engineering by numerous organizations and institutions.

David N. Crosthwait Jr. was an important figure in the history of engineering and his achievements had a lasting impact on the world. His work helped to improve the comfort and energy efficiency of buildings and continues to be recognized as an important milestone in the history of science and technology.

Elijah McCoy

Field of RailRoad Railroad Engineer

1843 - 1929

Elijah McCoy was a very smart and innovative man. He was born in Canada in 1844 and later moved to the United States. Elijah grew up to be an engineer and inventor, and he spent his life creating new and useful inventions.

One of Elijah's most famous inventions was a machine that could oil trains while they were moving. Before Elijah's invention, trains had to stop every few hundred miles to be oiled by hand, which was very time-consuming. But with Elijah's machine, trains could be oiled on the go, saving lots of time and money.

Elijah's inventions were so popular and useful that people started saying "the real McCoy" when they wanted to describe something that was the best of its kind. Even today, Elijah is remembered as a brilliant inventor and a pioneer in the field of engineering.

Charles B. Brooks

Field of Self-Propelled Street Sweeping Truck

Born-Died Year

Charles B. Brooks was a brilliant inventor from Virginia who made a huge impact on the world with his innovative ideas. Born in 1865, Charles was always fascinated by technology and spent his free time tinkering with machines and trying to come up with new ways to make things better.

One of Charles' most notable inventions was a self-propelled street sweeping truck, which was a revolutionary way to keep streets clean. The truck had brushes attached to the front fender that could be swapped out depending on the weather, so it could be used to clear snow as well as sweep away dirt and debris. Charles received a patent for his invention in 1896, and it quickly became a popular choice for cities all over the country.

Charles wasn't just an inventor, he was also an entrepreneur. He secured funding to produce his street sweepers and set up a factory in Scranton, Pennsylvania to manufacture them. The sweepers were so successful that the state government even gave Charles a $100,000 contract to produce more.

But Charles wasn't done yet. He also received a patent for a ticket punch in 1893, which was an early version of a paper punch. His punch was unique because it had a built-in receptacle to catch the round pieces of waste paper, which helped to prevent littering.

Charles B. Brooks was an incredible inventor who made a lasting impact on the world with his innovative ideas and inventions. His work inspired other inventors and helped to make our world a cleaner and more efficient place.

Phillip B. Downing
Field of Communication
1857 - 1934

Phillip B. Downing was an African American inventor who made significant contributions to the field of communication. Downing was born in New York and received a degree in engineering. He worked as a research scientist before becoming an inventor, and he received a number of patents for his inventions.

One of Downing's most notable inventions was the mailbox, which is a device that is used to securely deliver and receive mail. Downing's mailbox was an important innovation that greatly improved the efficiency and convenience of communication. Prior to the development of the mailbox, people had to rely on messengers or personal delivery to send and receive mail, which was slow and unreliable. The mailbox made it possible to send and receive mail quickly and securely, and it is still in use today.

In addition to his work on the mailbox, Downing also developed a number of other inventions, including an improved telegraph system and a device for measuring the speed of sound. His inventions had a major impact on the world, as they helped to improve the efficiency and reliability of communication and made many tasks easier and more convenient.

Downing's legacy as an inventor is a great source of inspiration for African American children, as it demonstrates the importance of hard work, curiosity, and a passion for learning. His contributions to the field of communication have had a lasting impact and continue to be used widely today.

Benjamin Montgomery

Field of Propeller for Steamboats

1819 - 1877

Step back in time to 1819 and join us in celebrating the birth of a true visionary - Benjamin Montgomery! Born a slave in Loudon, Virginia, young Benjamin's life was forever changed when he was purchased by none other than the future president of the Confederacy, Jefferson Davis. But this wasn't just any ordinary purchase - Davis made Benjamin walk almost 1,000 miles hitched to a horse while he rode in his carriage. Can you imagine the strength and determination it must have taken to survive that journey?

Despite the harsh start to his life as a slave, Benjamin was a gifted inventor with a unique understanding of how things worked. With the help of the books in Davis' library, he learned to read and even draft architectural plans. He quickly rose through the ranks and became the general manager of the Davis plantation, where he put his talents to work improving machinery and building newstructures.

But his greatest invention was yet to come. In 1847, he created a new type of propeller for steamboats that could navigate the shallow and dangerous waters around the plantations with ease. The Davis brothers were impressed and wanted to patent the invention, but due to a Supreme Court decision that stated slaves were property, they couldn't. But that didn't stop Benjamin! He continued to develop his ideas, even after the Civil War broke out and the country was in turmoil. Finally, in 1884, Benjamin received credit for his invention and exhibited it at the Chicago World's Fair and the Southern Exposition in Atlanta, Georgia. Benjamin's story is more than just an inventor, it's a story of resilience, creativity, and unwavering spirit. His legacy continuesto inspire generations.

Samuel Raymond Scottron
Field of Mirror Bracket
1843 - 1905

Samuel Raymond Scottron, is a brilliant African-American inventor from Brooklyn, New York! He was born in Philadelphia in 1841 and from a young age, he was fascinated with how things worked. He would take apart gadgets and toys just to see what made them tick.

As a child, Samuel moved with his family to New York City where he attended grammar school and even served as the sutler for the 3rd United States Colored Infantry during the Civil War. Despite the challenges he faced, Samuel never gave up on his dreams. He saved up enough money to attend Cooper Union where he studied engineering and graduated with flying colors.

After graduation, Samuel became an inventor with a mission. He devoted himself to creating inventions that would improve people's lives. His first invention, the special mirror bracket, was a huge hit and it allowed people to see themselves just as others saw them. But that was just the beginning. He went on to invent the curtain rod, and many other things that made life easier.

But Samuel was not just an inventor, he was also a true leader in his community. He worked tirelessly to promote racial harmony and fairness, and he was a member of the Brooklyn Board of Education. He even helped to fight for the end of slavery in Cuba and Puerto Rico. Samuel Scottron was a true revolutionary, who changed the world one invention at a time, a true inspiration for all the young inventor out there.

Augustus Jackson
Field of Food Science and Culinary Arts
1808 - 1852

Augustus Jackson was an African American inventor and chef who made significant contributions to the field of food science and culinary arts. Jackson was born in Philadelphia, Pennsylvania and worked as a chef and confectioner before becoming an inventor. He received a number of patents for his inventions, including one for a process for making ice cream.

Jackson's process for making ice cream involved using a machine to churn the mixture of cream, sugar, and flavorings, which helped to create a smooth and creamy texture. This process greatly improved the quality and consistency of ice cream, and it made it possible to produce large quantities of the dessert efficiently.

In addition to his work on ice cream, Jackson also developed a number of other food products, including a type of chocolate that was used to make chocolate covered ice cream cones and a process for making a type of caramel called "panacotta." His inventions had a major impact on the world, as they helped to improve the quality and variety of food products and made many tasks easier and more convenient.

Jackson's legacy as an inventor and chef is a great source of inspiration for African American children, as it demonstrates the importance of hard work, creativity, and a passion for learning. His contributions to the field of food science and culinary arts have had a lasting impact and continue to be used widely today.

Valerie Thomas
Field of Food Science and Culinary Arts
1943 -

Valerie Thomas, is a brilliant African American inventor who revolutionized the world of visual technology and earned the title "Mother of 3D Animation!" Valerie was born in 1943 in Maryland and was always curious about the world around her. She went on to study physics and mathematics, but her true passion was for invention and innovation. She began her career in the 1970s as a data analyst at NASA's Goddard Space Flight Center, where she quickly made a name for herself as a brilliant scientist.

Despite facing discrimination and barriers because of her race and gender, Valerie refused to let that stop her from achieving greatness. She was determined to make a difference, and she did just that with her groundbreaking invention: the illusion transmitter. This amazing device uses mirrors to reflect light in a special way, creating the illusion of a 3-dimensional image on a flat surface. Can you imagine watching a movie or playing a video game and feeling like you're right in the middle of the action? That's exactly what Valerie's invention made possible!

Valerie's invention was patented in 1980, and it has had a huge impact on the world of visual technology. It's been used in everything from medicine to movies and television, and it's still used today to create stunning 3D images. Because of her invention, Valerie earned the title "Mother of 3D Animation"

Valerie's story is truly amazing. Her relentless determination, her brilliant mind and her unwavering spirit make her a true inspiration for kids everywhere who want to pursue a career in science and technology. Her invention is a perfect example of how with hard work, dedication and a passion for innovation, you can change the world.

Sarah Boone

Field of Ironing Board
1832 - 1904

Sarah Boone, the brilliant African American inventor who revolutionized the world of ironing!

Born in 1832, in North Carolina, Sarah had a keen eye for design and saw a problem that needed solving in the household essential, the ironing board. The traditional ironing boards were small, poorly designed and made ironing difficult, especially when it came to ironing shirts and large clothing items.

Determined to make a difference, Sarah set out to design a better ironing board. And she did just that! She came up with a revolutionary design that included an inclined surface and a curved end, making it possible to easily iron sleeves and pant legs. She was awarded a patent for her invention in 1892.

Sarah's improved ironing board design quickly became the standard and it was widely used in households across America. It made ironing much more efficient and comfortable, transforming a tedious chore into a breeze.

Sarah Boone was a true game-changer, a trailblazer for black inventors, and her legacy continues to inspire future generations. Her creativity, skill, and determination to solve a problem, made a real impact on the world. Thanks to her, ironing has never been more fun!

Patricia Bath
Field of Eye Care
1943 - 2019

Patricia Bath was an African American ophthalmologist and inventor who made significant contributions to the field of eye care. Born in 1942 in New York, she was the first African American woman to complete a residency in ophthalmology and the first African American woman doctor to receive a medical patent.

One of her most notable inventions is the Laserphaco Probe, a device that uses a laser to remove cataracts, which is one of the leading causes of blindness. The device is a major improvement over traditional cataract surgery which uses manual tools that can be less precise and riskier for patients. Patricia's invention allows for more accurate and less invasive cataract removal, improving patient outcomes and recovery time.

Dr. Bath's pioneering work, not only made a significant impact in the field of Ophthalmology, but also as a woman and as a person of color in a field that was, and still, is dominated by white men. She was a trailblazer, who shattered barriers and inspired many in the field.

Patricia Bath is a true inspiration, her dedication to improve people's lives and her pioneering spirit, serves as an example of what can be achieved through hard work, determination, and a passion for innovation.

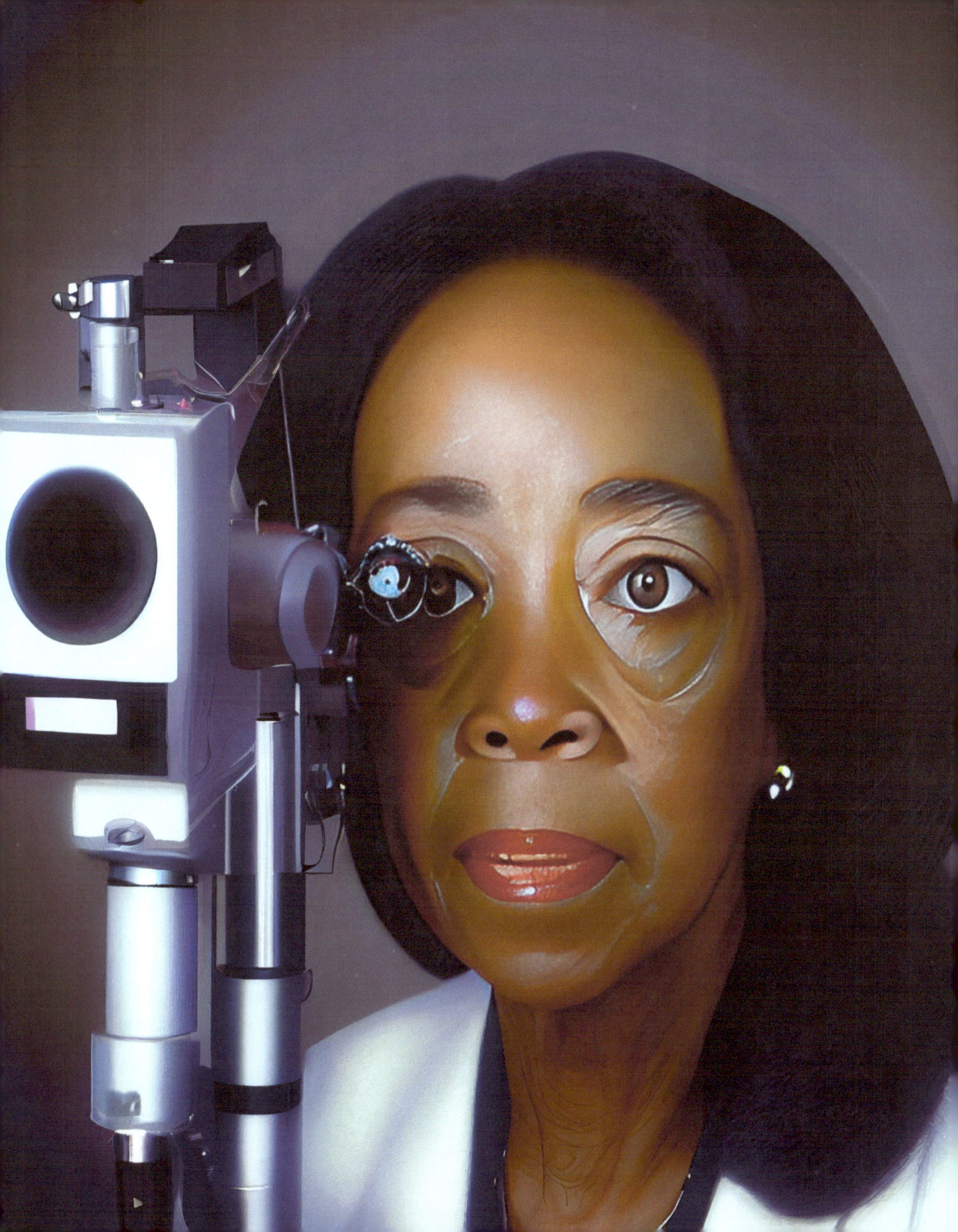

Madam C.J Walker

Field of Hair & Beauty
1867 - 1919

Madam C.J. Walker was an African American entrepreneur, philanthropist, and civil rights activist who made history and changed the game for black women in business. Born in 1867 as Sarah Breedlove, she rose from poverty and struggle to become the first female self-made millionaire in America, breaking barriers and proving that with hard work, determination, and the right idea, anyone can achieve success.

It all started with a personal problem, Madam Walker was experiencing hair loss and scalp issues, she tried every hair care product available but none of them worked for her. That's when she decided to take matters into her own hands and began experimenting with different hair care formulas. Her hard work paid off when she finally created her own line of hair care products specifically for black women, and began marketing them door-to-door. Her business quickly took off, and before long, she was opening her own factory, beauty school, and chain of beauty salons. Her company, Madam C.J. Walker Manufacturing Company, became one of the first black-owned companies to be listed on the New York Stock Exchange. She was not only a successful businesswoman but also a philanthropist, she supported various causes such as education and civil rights.

Madam C.J. Walker's story is an inspiration to many women, especially those of color. She broke barriers, shattered glass ceilings and proved that anything is possible with hard work and determination. Her legacy continues to inspire generations of entrepreneurs and business leaders to this day, showing that with perseverance and grit, you can make your wildest dreams a reality.

In conclusion, the contributions of black inventors to technology and society are immeasurable and often overlooked. From the creation of the traffic light to the development of the ironing board, these individuals have played a vital role in shaping the world we live in today. Their innovations have improved the lives of millions and continue to inspire future generations of inventors. It is important to acknowledge and celebrate the accomplishments of black inventors, not only for their contributions to society but also to serve as a reminder that anyone, regardless of their background, can make a significant impact on the world.

www.ingramcontent.com/pod-product-compliance
Lightning Source LLC
Chambersburg PA
CBHW051209220526
45473CB00003B/967